SpringerBriefs in Computer

Series Editors

Stan Zdonik
Peng Ning
Shashi Shekhar
Jonathan Katz
Xindong Wu
Lakhmi C. Jain
David Padua
Xuemin Shen
Borko Furht
V. S. Subrahmanian
Martial Hebert
Katsushi Ikeuchi
Bruno Siciliano

For further volumes:
http://www.springer.com/series/10028

Charles J. Petrie

Automated Configuration
Problem Solving

 Springer

Charles J. Petrie
Stanford University
Bastrop, TX
USA

ISSN 2191-5768   ISSN 2191-5776   (electronic)
ISBN 978-1-4614-4531-9   ISBN 978-1-4614-4532-6   (eBook)
DOI 10.1007/978-1-4614-4532-6
Springer New York Heidelberg Dordrecht London

Library of Congress Control Number: 2012940242

Printed on acid-free paper

Springer is part of Springer Science+Business Media (www.springer.com)

Preface

Automated Configuration has long been the subject of intensive research, especially in Artificial Intelligence, as it is a pervasive problem to be solved and is a good test of various knowledge representation and reasoning techniques. The problem itself shows up in applications as various as electrical circuit design, utility computing, and even concurrent engineering. We define this ubiquitous problem and show the various solution techniques.

We survey about 20 years of these techniques, from the mid-1970s until the mid-1990s. During this time, various general approaches were developed, as well as more specialized techniques. We survey the development of the general problem solving techniques for automated configuration, based on both published academic work and patents.

This book is intended to be an introduction to the topic and a gateway to more detailed descriptions of configuration technology while presenting a possibly different perspective in some regards and covering previously overlooked material, especially the commercial development of configuration technology. This was discovered largely not in the course of academic research but during patent litigation that not only informed the author about a large set of patents but also about the commercial worth of configuration problem solving.

The author depended heavily upon a shorter survey [71]. For more detail on how constraint solving technologies are used for configuration problem solving, Junker [27] is very thorough and recent. For an industrial research perspective, see [23], which adds some detail, especially about specific configurators and specific logics used for problem description and solving.

The author thanks Sudhir Agrawal at the KIT and Ulrich Junker at IBM for their careful reading, comments, and corrections; and Jim Batchelder and Victoria Smith for their guidance and support during the patent litigation that provided the opportunity for this review and analysis.

Karlsruhe, Germany, April 2012   Charles J. Petrie

Contents

Chapter 1
Configuration Problem Definition and Technology

Abstract This chapter provides an overview of automated configuration problem solving, including an definition within the context of design and AI Planning, and special cases.

1.1 Overview

People have been designing and engineering artifacts, such as bridges, steam engines, and airplanes, for a long time. As computer scientists gained more experience in the late twentieth century with designing computers and electronic circuits, it became clear that we could use computers to help us perform such design. In fact, in some cases, the design could be automated.

Such automated design work became, quietly, a much more successful result from the field of Artificial Intelligence (AI) than was the more touted field of "expert systems" [15], though that branch finally showed real results too. Both areas of AI were based upon symbolic representation of knowledge and computational logic: inferring by some kind of proof mechanism the solution to a given problem [22].

The focus and energy of AI in the twenty-first century is based largely upon statistics and probabilistic reasoning: this is the foundation of today's driver-less vehicles and other kinds of robots. However, symbolic representation and reasoning was perfectly suited for a large set of design problems as the results would define a machine that should work. A major advantage of computational logic is that synthesizing the design of a machine by proof means that the design is provably correct. If the designer has correctly understood and represented all of the constraints governing the machine, the design is as correct as is possible to know until it is actually built.

A major limitation of this approach is that unlimited design may require the development of novel features. This is possible with AI but beyond the capabilities of currently well-understood science and engineering. We do know how to use AI technology to support people in developing novel design by coordinating their indi-

C. J. Petrie, *Automated Configuration Problem Solving*, SpringerBriefs in Computer Science,
DOI: 10.1007/978-1-4614-4532-6_1, © The Author(s) 2012

vidual decisions [53]. But creating novel designs routinely requires more technology development than has occurred to date.

Fortunately, there is an important subclass of design called "configuration" that is a fundamentally solved problem. Configuration problems are those where the goal is to design something new from old components: parts that we already have developed and have as many to use in the design as may be required.

Examples of configuration are designing: an electronic circuit from standard components, a cloud/utility computing system that can handle certain peak loads and applications, new models of automobiles and laser printers from an existing parts catalog, new factories based on existing manufacturing technologies, and mass transit systems based upon currently manufactured rail and moving equipment. Furthermore, much of the planning for building such artifacts can also be expressed as configuration problems where individual actions are the "parts" to be configured, and which lead to the final product.

Configuration is thus a very important and large subset of the general design problem. The fundamental work on how to automate the problem solving of configuration began in the mid-seventies and was largely complete by the mid-nineties. After this time, most, if not all, of the new developments were for specialized cases of configuration, which are of course in themselves important.

This work will review the important developments and technologies for automated configuration during this time when most of the problems were solved using varieties of symbolic representation and reasoning. The text is intended for the student of computer science who may want to understand better these technologies either to solve new problems or just be better educated about this important subtopic of AI, which continues to be important.

1.2 Configuration Problem Definition

A configuration is an ordered arrangement of a set of parts. A solution to a configuration problem is the selection and arrangement of a set of parts that satisfy the problem requirements as well as constraints associated with these objects [71]. The configuration solution is a specific instance of all possible configurations. Parts to be configured may be components or instances of those components depending upon how specific the requirements are.

Parts may be instances of component types, or, in the simplest cases, there may be concrete objects, such as in a jigsaw puzzle. In other cases, the parts are types of components, such as configuring a car according to a customer's wishes from a set of possible components. One chooses a type wheel, for example. The concrete objects may be instances of a component type, such as a Lego(TM) part, or an actual wheel that one purchases at a shop. There may be an unlimited number of parts for a given configuration problem: there may be a limited number of lego part types, but one can order as many instances of each type as you wish for your construction project.

Sometimes component types are simply referred to as "components" and to all of the components and possibly their instances as "elements" of the configuration. Here, we use the word "parts" to mean any of these. The configuration problem is to find a configuration that satisfies some set of requirements, given a set of all possible parts and their characteristics, including constraints on their arrangement.

A configuration may be an arrangement of components that can be instantiated at manufacturing time. For the purposes of design, these components suffice and are usually endpoints in a graph of components, or "leaf nodes", as there are no other parts of which these are types. For instance, alloy wheel model number 2938 is a type of alloy wheel which is a type of wheel, but there is no further specialization of this part. There will be a concrete instance of such a wheel at manufacturing time, but at design time, the configuration is complete.

Configuration may concern any set of components, including computer software and hardware. Most often in the literature, mechanical design is considered, though some of the earliest work concerned electrical circuits.

Configuration problem solving is a specialization of the general design problem, in which no new types of components may be added to the problem but the number of instances of a component type might be unbounded and perhaps not all possible component types are used in a given configuration [7, 21, 71]. Configuration design is also called "Class 3 Design" in [6]. In a general design problem, new components may be designed as part of the problem solving. Designing a new model of car is often not a configuration problem as new components may need to be developed: e.g., the new car may require a new alloy wheel to be designed that did not previously exist in the catalog of parts.

Configuration has long been an object of study within the symbolic reasoning segment of the Artificial Intelligence (AI) community because some problems defy conventional techniques. However, it should be noted that configuration is an old problem that predated computers: steam locomotive designers, for instance, often solved very complex design problems without computers. As computers have become more common, they have been used to solve simple configuration problems without what would be considered AI. For example, even in the 1970s, it was common to use small computers to input the parameters of a house and select, from a stored catalog, the model of the available air conditioners that would be the right size for the house.

As computers became more complex, enabling symbolic reasoning, AI researchers became interested in larger sets of configuration design problems such as electronic circuits, telecommunication systems, and even computers themselves. Additionally, "planning" is often considered as a configuration design task in which some final state, such as an arrangement of stacked blocks, should be achieved [71]. "Scheduling" is one kind of such planning design problems where the components are often task assignments to time periods under various constraints. Travel planning and shop floor scheduling [67] are examples of such configuration problems. Another specialization is "space planning" in which the components are objects and the constraints are all spatial [3].

Component instances may have associated parameter values that are determined at configuration design time, such as the size of a printer roller. To the degree that these parameters can vary, problem solving is more difficult and can approach the general design problem as they may, in extreme cases, effectively define new components. Even a simple water tank becomes a new design object when it is expanded from in volume three orders of magnitude. When component properties are fixed, the problem is simpler.

Chapter 2
Configuration Models

Abstract This chapter provides an overview of the models used to represent the configuration problem and used for automated solving. The most common models augmenting rules and catalogs are classes, constraints, goals with restrictions, and resources. We also briefly describe Truth Maintenance Systems as a model that may be used in conjunction with these.

2.1 Rules and Catalogs

The object of computer-aided configuration is to automate as much as possible, using a computer program, such an arrangement of component instances. In order to do this, there must be a computer readable representation, a "model", of the components, the constraints, and the requirements. Were there such model, a computer could, with an algorithm that has not yet been invented, solve the configuration problem of arranging the letters of the alphabet into this text of this book. A model is necessary, but not sufficient, for automatic configuration problem solving.

One dimension of the way that configuration technologies differ is in the type of model. That is, when we describe a configuration problem solving technology, it is useful to distinguish it from others by its different features. One of the most important ways to distinguish configuration technology is by the kind of model a technology employs.

Any model of a configuration problem must contain at least the parts to be arranged as elements. This is typically a database, or catalog, of components to be selected. If we have only a finite set of actual Lego® blocks to assemble, our model consists of just these parts (types of blocks).

If we model lego block types so that a computer can reason about how to assemble together instances of them, then this is a component model that has at least two levels: a set of types and as many instances of each type as we need to configure the artifact. If we are given user requirements that refer only to the aggregate properties of such

C. J. Petrie, *Automated Configuration Problem Solving*, SpringerBriefs in Computer Science,   5
DOI: 10.1007/978-1-4614-4532-6_2, © The Author(s) 2012

a configuration (e.g., size and shape of the boat), such a model suffices for automatic configuration with some algorithms.

In this case, it may suffice to use only rules to describe our lego blocks. We might say, for example, that all "short blocks" have three pegs/holes and all "long blocks" have five pegs/holes, and all blocks have a width of 2 pegs. Then we could write rules about how to combine such blocks: e.g., to attach one block to another, there should be at least one peg on one block and one hole on the other that are free to be used for the attachment. The block descriptions would be a sufficient model for some configuration engine to solve a given problem. An additional simple model would be to create a catalog or database of available parts.

2.2 Class Hierarchies

For other than very simple problems, a completely rule-based approach with a parts catalog is inadequate, if only for reasons of programming complexity. The user requirements may need to be stated in terms of larger components and/or aggregate functional requirements. For example, we may want to configure an automobile that can go 120 mph. Because we want it to go that fast, we know that the properties of the sub-components, for example the wheels and engine, will need to provide and support such speed. Such requirements tend to be of two kinds. One is a requirement for subcomponents: e.g., a computer should have a CPU, memory, and an I/O bus. Another is for functionality or capacity: e.g., the computer should have at least 1 TB of storage. Both such requirements will lead to inclusion of new subcomponents in the configuration.

In such cases a more complicated model using class hierarchies is useful so that we can easily compute the properties of the design components, possibly through inheritance. Other relations may also provide additional guidance for automated problem-solving. The use of such a model is often called "model-based" configuration but it is so common that almost all approaches that perform complex configuration have some kind of class model, even if embedded in rules.

In a class hierarchy, subclasses descend from superclasses. Each element of such a decension tree is a class or an instance of a class. Each subclass will have at least one superclass. This is commonly called a "is-a" relationship. Subclasses are specializations of their superclasses. Members of any subclass inherit the attributes, or slots, of a superclass from which it is descended. They have further attributes make the subclass a specialization of the superclass. So "Jaguar" is a subclass of "Automobile" that has more specific attributes than does a general automobile. The component "engine" may have specializations of "gasoline engine" and "diesel engine".

Complete class hierarchies must include a class layer of components which have no specializations or sub-components. These are often called "primitive", or "leaf", components. A configuration will consist of instances of these primitive components. For example, the class "BMW Model 531e" may have no further subclasses.

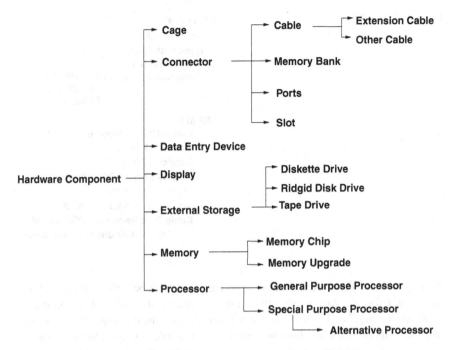

Fig. 2.1 COSSACK class hierarchy

The design configuration would be complete with such classes, but a model could contain class instances, such as "VIN930bo193x0222".

It is difficult to track down the earliest instances of such class hierarchies, but certainly one of the earliest famous cases is that of the general (domain independent) expert system shell EMYCIN [68] that included a hierarchical structure called a "context tree". Each object in this structure was of a particular type, associated with certain parameters. Parameters were inherited from parents (superclasses) and there could be instances of each type.

A good example of such a class hierarchy is shown in Fig. 2.1 which is a simplification of the hierarchy from COSSACK [20]. The top level superclass is "Hardware-Component". Subclasses of this top level component class include "Connector" and "Cage". Connectors include the subclasses "Slot", "Ports", and "Cable". "Cage", which is a type of component that may contain other components, is a primitive, or leaf, component as it has no subclasses.

Even some of the earliest rule-based systems used some simple class hierarchies. For example, Fig. 2.2 shows a simplified example of R1/XCON component descriptions [39].

Class, or "frame", systems are very specific object-oriented technologies that predated the mainstream use of object-oriented programming (OOP) [60] but inspired it. In frame systems, the objects are data types that inherit from one another. In OOP,

RK711–EA
  Class: Bundle
  Type: Disk Drive
  Supported: Yes
  Component List: 1 070–12292–25
                                    1 RK07–EA
                                    1 RK611

RK611*
  Class: Unibus Module
  Type: Disk Drive
  Supported: Yes
  Priority Level: Buffered NPR
  Transfer Rate: 212
  Number of System Units: 2
  Cable Type Required: 1070–12292
                         from a disk drive unibus device

the objects are software components with datatypes with behaviors, which can be inherited. However, they are two distinct technologies that should not be confused.

As shown in [2] and [69], frame class hierarchies and the properties of the classes and their inheritance could be represented in logical rules. As these became understood as static structures, it was understood that it was more efficient to use static specialized frame representation systems such as those used in Knowledge Craft, KEE [29], and the MCC "Proteus" [46] framework that combined forward and backward-chaining rules, truth maintenance, and frames. Such frame/class systems had attributes, often called "slots", associated with each class (not to be confused with the common domain-specific class "slot" referring to a specific location in the artifact where a component may be inserted).

Later systems use other types of classification subsumption systems that have equivalent but perhaps more efficient mechanisms, such as the use of description logic [41] in PLAKON [10, 27].

Whether done in rules or more efficient frame/slot systems or possibly even more efficient description logic, constraints could be placed upon the values of attributes of each class, such as cardinality. For instance, an "automobile" might be a class with the attribute "wheels" and the cardinality of this attribute might be restricted to the number four. Such class systems are frequently also called "ontologies" as the they are not simple taxonomies but more complex models that constrain the use of the elements in the models. (Ontologies that use various forms of computational logic [22] are used today to define formal semantics for various terms on the Wold Wide Web.)

In addition to the "is-a" relation, classes may have arbitrary relations with respect to each other. One that is very useful for configurations is the "part-of", sometimes called "child-of" or, inversely, "has-part", relation that describes sub-components [2] that are necessarily a sub-component of a super-component. This is an alternative to using the attributes of a class to constrain the class instantiations.

For example, instead of defining the class "automobile" as having an attribute of "wheels", this a could alternatively be defined as another class with a "part-of" relationship to automobile. To elaborate the example, the class "automobile" might be defined to always have a "has-part"/"part-of" relationship to the classes "engine", "chassis", and "wheel". In addition, constraints are often stated that an automobile should contain one each of these first two classes and four of the third. Other types of common relationships are "connects" and "contains". Similarly, an "engine" may require the sub-component "piston". A valid configuration respecting such a model would always result in car designs that have one engine, one chassis, and four wheels, and the engine would have at least one piston.

Any given model may have any finite number of such pre-defined relationships in order to guide problem solving. The semantics of the relationships and constraints is ultimately defined by the problem-solving algorithm that uses them. Some simple configuration systems force such sub-components selection [21].

In general, a configuration may not necessarily include component instances of all sub-components of a component. More general configuration systems may represent sub-components as optional parts. For example, a "computer" may have a sub-component of "optical disk reader" but this component is optional. Or it may be that some kind of optical disk reader is required but this might be either a "CDROM" or a "DVD" reader. If no subclass was defined, then either one, but at least one, must be part of the configuration. In this case the classes would not represent mandatory sub-components but there would be a rule that at least one instance of one of these subclasses would be in the configuration. For example, we don't know in the beginning of the configuration problem whether this type of computer will have a CDROM or a DVD but we know it will have one or the other.

In addition, some configuration systems use a functional hierarchy that uses the is-a hierarchy to inherit functional properties, which has some advantages in satisfying functional requirements and constraints [4, 21]. This is similar to OOP in that artifact functions and software behaviors are at least analogous.

2.3 Constraints

A constraint is a very general and useful formalism. It is often expressed as a rule about the combinations of the assignments of values to variables. A common example is map coloring: "any two adjacent countries must have different colors".

A common way to express this is as a triple: "(Color Germany Red)" would mean that he color of the map element "Germany" has been assigned the "color" "red". However, for constraint representation, it is convenient to use a single variable name, such as "Cgerm". Then color of any country on the map is expressed as the value of a variable. So, in addition to "Cgerm", the color variable of Austria might be called "Caus", and that of Switzerland might be Cswiss. We may have only the color values of "blue", "red", "green", "white", and "black" to assign to these country color variables. If we assign "blue" to "Cgerm", then "Caus" can only be one of the

other three colors because of the constraint. Assigning one, say "red", to "Caus",
means that "Cswiss" must be either "white" or "black".

To get around the awkward variable names, sometimes variables are abbreviated
with a particular syntax, such as "Germany:Color". Then our constraint might be
expressed as

$$\neg \wedge (?X:Color = ?Y:Color)(Adjacent?X?Y).$$

This relates to our previous discussion of classes because class have attributes,
also called "properties". "Color:Automobile" would be the variable relating to the
class Automobile. An assignment this high in the class hierarchy is typically a default
that propagates downward. For example, "(Color Automobile Black)" may be used
to mean that all automobiles (in this model) are colored black.

Constraints are conditions over the assignment of values to the properties of
components in a configuration that allow only some to be valid. The constraints may
be very specific to the values assigned to a slot on a frame or more general in how
components can be combined. A simple example of a constraint is that an automobile
may have no more than four wheels.

Constraints are a common part of models used to represent and solve a configuration problem. In some systems, only constraints are used. That is, constraints may
only be part of a class model, or they may be the only model used. We discuss the
latter in Sect. 3.2.

The constraints associated with configuration design problems often include constraints on how instances of one component type are connected to another, often
with an instance of a third kind of component called a "connector", where some
constraints determine that particular components can be connected only at special
points called "terminals", which may be "slots" (distinct from frame slots) and/or
"ports" [1, 20, 21]. Typically a "slot" is a terminal into which some component is
inserted rather than just connected to, as with a "port". Typically, the restriction of
port and slot connection points also includes a new requirement for a particular class
of connector, such as a USB cable.

All constraints, including connection constraints, may be associated with components at any level of the specialization hierarchy tree. All components lower in
the tree inherit these constraints from components higher in the tree. For example,
every "disk drive" must be connected to a "system bus" at a "bus interface port". So
specialization disk drive "160 GB-drive" component must be connected at a specialization of such a port by some specialization of such a bus, which is a specialization
of "connector".

A related common type of constraint associated with configuration design problems includes spatial constraints. Some components must have a particular spatial
relationship with others. This may involve being nearby or being physically inserted
into each other, such as a board into a slot. Typically there are restrictions on how
many and which components can be inserted into the slots of another component.

Some of these constraints represent physical constraints (no more than six power
adapters of any kind may fit on the power strip and no more than three adapters

larger than one inch may fit) and some have to do with functional requirements such as power (the total power requirements of all power adapters on the strip may not exceed 500 W).

Constraints may be represented as rules apart from a class system, and/or as constraints associated with the component classes and/or their slots. In either case, they are logically equivalent and express relations constraining the valid configuration.

2.4 Constraints as Goals and Restrictions

In some systems, all conditions on components are called "constraints". The solver simply finds value assignments to the variables that satisfy all of the constraints, possibly with the aid of class models and other types of knowledge. However, it is often useful to distinguish between different kinds of constraints. Such a distinction is particularly useful for Generate and Test problem solving methods, discussed in Sect. 3.3. These methods will typically use a constraint and class model discussed above, but with further distinctions.

Some constraints are restrictions on how components existing within a given configuration may be combined: these are compatibility constraints. For instance, such a constraint restriction might be that a convertible may not have a hard top. If during the configuration synthesis of a convertible car, a hard top is added to the configuration, then this constraint is violated. In order to fix a constraint violation, we must either remove the hard top from the configuration or remove the selection that the car is to be a convertible. Connection, physical, and spatial constraints are other types of restrictions.

Some constraints may restrict the aggregate of resource consumption. A typical type of restriction constraint is a budget of some kind: the aggregate of components added to the configuration cost too much money, weight, or some other budgeted quantity, including perhaps time to build. It is often not possible to fix such a violation by simply adding something else to the configuration. A notable exception are resources, such as money and electrical power. For example, it may be possible to add an additional power supply. If this is not seen as replacing one power supply with another, then the available power is just a constraint and so the constraint itself is changed (or this is represented as a compatibility constraint). So the solution is either to change an earlier variable assignment or to change the constraints.

Other types of general constraints are requirements, or requests, that some additional components or statements about the design should be added to the configuration, if these requests are not already fulfilled. If we have selected the car to be a luxury touring car, then there may be an added requirement that a multimedia package be added to the car. This is a request that can be satisfied by adding such a package to the configuration if we have not already done so.

So far, we have discussed restriction and requirement constraints, Both can be handled by constraint solving systems [27]. However, there is a difference between the them that can be exploited by a problem-solving system. As discussed above, typically the way that constraint violations of the first type is solved is to change a previous selection of variable assignment. Such constraints are not violated unless such selections are made. However, the requirement constraint may be initially violated until some selection is made. The same is true for resource requirements, such as a requirement for disk capacity in a computer. The more general notion of such requirements is that of a *goal* [50, 71].

While sometimes the choice between saying a condition is either a goal or a restriction is a representation decision, it is often useful to distinguish such requirement-type constraints that can be satisfied by making design decision from restriction-type constraints that are only violated when such decisions are made. One reason is that there are often many more possible restrictions to be considered than are tractable and proceeding by satisfying necessary goals until some restriction is violated often reduces the constraint problem space. In addition, the goal representation facilitates configuration task decomposition as it explicitly labels the expression as possibly leading to decomposition, distinguishing it from restrictions. A goal representation also ensures that there is a valid reason for each of the component assignments, thus providing an explanation and eliminating unneeded assignments that may have occurred as a by-product in pure constraint satisfaction.

The component instances in a correct, or valid, configuration have properties that collectively satisfy the top-level goals of the user do not violate the constraint restrictions that may be associated with the component properties. It may not be possible to satisfy all of the requirements and restrictions for a given problem, in which case the configuration will still consist of a set of component instances but will not be a correct configuration. Still, it may be the best that can be done because the problem is over-constrained. Also, it may be that constraint violations should be deferred for a while during problem solving but ultimately resolved producing a correct configuration [50, 71].

2.5 Resources

It is common in many configuration systems to represent some component properties as well as some user requirements as resources. Indeed, some configuration systems use only resources [25] for configuration problem solving.

Resources are commodities that may be both provided and consumed by components, and may be required by the user. An individual component may supply a unique resource: i.e., a particular CPU provides 10 GHz serial processing speed. Or it may supply a resource that aggregates. If the computer needs a total of 500 GB of disk space, this may be achieved by five 100 GB disk drives. If the two disk drives plus the CPU consume 100 W of power, then this power may be supplied by either a single power supply of 100 W or by configuring two 50 W power supplies that work together.

In such systems, there is typically no distinction made between requirements and restrictions: there are only resources needed and consumed. Such resources are often represented as properties of the individual components. The property of a disk drive may be that it consumes 50 W. The property of a power supply may be that it supplies 50 W. Resources may concern component aggregate properties. A user resource goal, which is really a restriction, may be that the total weight of the components be less than 2 lbs. A resource requirement-type constraint may be that if the total heat output is more than 30 W, then a fan must be configured. Resource-oriented systems treat all such resource constraints similarly. Indeed, all requirements and constraints are considered as resource provision or consumption that needs to be balanced. This is a unique view on configuration and though successful, has not been applied much since the original work [25]. Typically this resource computation is combined with other types of problem-solving [27].

2.6 Truth Maintenance

While they are only ever used in conjunction with other models and with specific reasoning systems, Truth Maintenance Systems [12, 16] are an important model for configurations systems. Essentially these systems provide a background support to ensure the consistency among some set of beliefs. The model is complex but the essential idea is that the reason for belief in some assertion should be supported by belief in some set of other assertions, or lack of belief in them.

The simplest use of a TMS is to record the results of firing a rule. For instance, if the rule is that

$$\wedge(?X : Daytype{=}Weekend)\neg(?X : Weather{=}rain) \Rightarrow (?X : Picnic = Yes)$$

then we can plan a picnic for a day on the weekend as long as we don't believe it rains on that day. Upon firing this rule for a particular day, say "Saturday", a TMS will add a new justification node for having a picnic on Saturday to a possibly already existing network of justifications. Figure 2.3 shows a simple example justification that might result from such a rule firing, with the belief that Saturday is a weekday somehow also justified. The *IN* and *OUT* node labels refer to whether a node has a valid justification: the former if so and the latter if not. A justification is valid if all of its supports marked with + are *IN* and all of its supports marked with − are *OUT*. Any node may have multiple justifications. See [50] for a further explanation.

The important point is that a TMS will automatically retract the belief in planning for a picnic on a day as soon as it is believed that it will rain that day. This functionality can be of use in tracking the validity of configuration choices.

However, the use of either the original TMS or the ATMS has long been understood to be problematic [48], because, if one simply ties a rule-firing to the production of a TMS justification structure, the propagation of change in the TMS network can be difficult to understand and predict. For instance, a top-level choice in the customization of an automobile as a sports car might lead to many choices, such as the choice of high-profile tires. If the top-level choice is revised, then all such

Fig. 2.3 A simple TMS
justification network

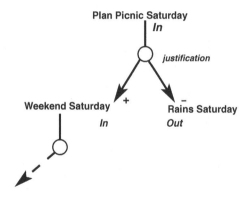

further choices including the choice of tires will be automatically revised as well, even though the tires may actually still be desired in the new sedan configuration. Given a top-level rule firing that is changed, all subsequent rule firings that lead from the consequent of that rule will be changed whether or not a domain model requires such change. In the worst case, work should be repeated and in the worse, interactive configuration will be confusing. A use of TMS in configuration is described in [50] and discussed in Sect. 4.3.

Chapter 3
Reasoning Techniques

Abstract This chapter discusses the two primary techniques for the solving of configuration problems: Constraint Satisfaction and Generate and Test. We also consider briefly specialized versions of Generate and Test, TMSs, and Case-based Reasoning as problem solving techniques. The following chapter considers specific systems in the context of the reasoning frameworks described in this chapter.

3.1 Beyond Models

A model is necessary but not sufficient for configuration problem solving. We cannot just give a computer a model and say "solve". We must tell the computer how to solve the problem by programming the computer with code that implements a particular algorithm understands the model and which will work in such a way as to solve the problem.

AI-based configuration systems typically use algorithms that reason with symbols, rather than statistics. algorithms are said to be "sound" if their solutions are necessarily valid, such as the ones derived by proof methods. They are said to be "complete" if they can find all possible valid solutions. Many of the algorithms for solving configuration problems are not complete, or sound, or neither, simply because otherwise the problems would take too long to solve in some cases.

There are two general methods used: generate-and-test and constraint solving. In addition, various hybrid versions and variations of these are used as well.

3.2 Constraint Satisfaction Problem Solving

A fundamental AI technique is the satisfying of constraints collectively in a finite variable Constraint Satisfaction Problem (CSP). In a CSP, there are a finite set of variables that must be assigned values. There is also a set of constraint relations that

C. J. Petrie, *Automated Configuration Problem Solving*, SpringerBriefs in Computer Science,   15
DOI: 10.1007/978-1-4614-4532-6_3, © The Author(s) 2012

determine legal combinations of values to variables. For a formal definition, see [58]. The variables, their possible values, and the constraints are defined prior to problem solving. That is, no new variables or constraints are typically added after problem solving begins. This corresponds to the requirement that no new component can be added during configuration.

Constraint Problem Solvers (CPSs) typically express all conditions on components as constraints, and typically these are finite domain variable constraints. This means that all of the features of the configuration that might be determined in creating a configuration are enumerated, and all of the possible values that might be determined for each variable are also enumerated, both in advance of problem solving, so they should be finite. This representation is adequate for many problems and can be solved by CPSs, as described below.

Other common approaches allow the variables to have an infinite number of possible assignments, typically real numbers. There are special algorithms for such problems, which may be very difficult to solve and they may not be considered configuration problems if they essentially require the design of new parts. For instance, the modification of an existing wind generator rotor for special wind conditions is likely to result in a almost completely new rotor, the design for which cannot be accomplished by straight-forward constraint satisfaction.

Standard CSP solving techniques, usually by some form of "k-consistency" [34], are also often characterized as unsound as they may produce an answer that still does not satisfy all constraints: i.e., no globally consistent solution has been found. The reason for this is that algorithms that ensure all constraints are satisfied take too long for computers to execute in the worst cases. Still, a candidate solution can be quickly verified to see whether it is correct, so that the combination of k-consistency and verification is sound. However, it may not be complete. If checking reveals an inconsistency (at least one constraint is violated), the method of backtracking determines completeness.

A TMS, as described in Sect. 2.6, can be used by itself for simple constraint solving. Figure 3.1 shows the constraint that only one of the three variables can be assigned the color *red* and also shows one of three consistent labelings of *IN* and *OUT* for this justification network (notice one node is shown three times). Obviously constraint formalisms provide a much more compact problem representation but the constraint propagation mechanism for solving such constraint problems is similar to that of finding a consistent labeling of a TMS network.

The original TMS [16] provides support for *dependency-directed backtracking* [62] as a way to resolve conflicts, such as constraint violations. This mechanism looks at all variable assignment values that have an assumption basis that lead to the current conflict creates an *AND/OR* tree of assumptions that can be retracted in order to resolve the conflict. Sometimes the assumptions are simply the potential assignments of possible values to a variable.

This is useful in conjunction with other reasoning mechanisms, such as rule firings and class inheritance that may produce such constraint violations. A simple example is that if we learn that we cannot plan a picnic on a weekend day, then, according to

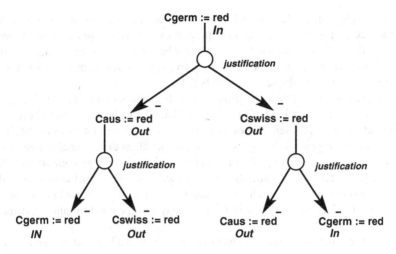

Fig. 3.1 A Constraint in a TMS

the rule that we plan to have a picnic unless it rains, we would abduce that it rains that day.

Obviously, such rules incorporating assumptions can be quite complex. One successful implementation using a TMS to perform abductive reasoning restricted the assumptions to those involved in designing custom chips [63]. But in general, simply representing all possible value assignments to variables as such assumptions and then backtracking, trying each one in turn, is not a good technique for constraint solving.

The later ATMS [12] sought to avoid backtracking by keeping track of which sets of beliefs were consistent. In the simplest case, a belief could be that some value should be assigned to a variable and consistency is determined by constraints. At least one commercial configuration system is known to have used an ATMS successfully in such a manner though with some knowledge-based guidance [24].

Such pure constraint-based systems typically represent both the user requirements as constraints as well as all conditions on the components. A configuration may result from applying a variety of constraint solving techniques, many of which produce all consistent solutions and fail when no globally consistent solution may be found.

Constraint satisfaction techniques are applicable to a specialization of the configuration design problem when all of the components and their properties, including how they may be assembled and connected, are known ahead of time and only their spatial arrangement and/or parameters associated with the components need be assigned values in order to satisfy the constraints [71]. Such configuration problems can be called "assembly" problems: e.g., construct a boat out exactly five provided lego blocks. While the solution of such CSPs is a subject in its own right, most configuration systems address the problems that cannot be easily formulated as CSPs: otherwise the problem may be solved with one of various CPSs available.

CSP solving difficulty roughly increases as the constraints cover more variables, corresponding to component properties and user requirements. In many cases, whether user requirements are met cannot be determined solely from the properties of any individual component but only from the aggregate properties (such as collective weight) of multiple components [71].

Thus, even some CSPs can be difficult to solve without some knowledge about how to solve a particular kind of problem, and knowledge becomes more necessary as the configuration problem moves away from an easy formulation as a CSP. This knowledge can be general (it is good to first assign values to the variable associated with the most constraints) or problem-domain-specific (first assign memory to CPUs before disk drives). Such knowledge, often called heuristic knowledge, guides the problem solver so that the search for a correct configuration does not become infeasible. Class models, as described in Chap. 2, may also be used to guide constraint solving.

A simple configuration problem would be to construct a boat out of a given set of lego blocks, perhaps not using all of them. In such a case, we say that the lego blocks are the actual component instances. If there are not too many possibilities, a CPS can be used. But even when the number of possible component instances is finite, the search space may be so large that enumeration of all of the potential component instances, their possible connections, and possible physical arrangements is too large to be feasible.

However, class models introduce the notion that not all possible parts and constraints may be applicable to a particular solution. In these difficult configuration design problems, it cannot be determined ahead of time which components, and how many instances of which, will be used in the configuration.

In worst cases the enumeration of possible configurations would be countably infinite because component instances can be generated as needed without bound, impossible with finite-domain CSP solving. An example is to construct a boat to specifications out of any number of instances of types of lego blocks. In this case, the lego blocks are component types and we generate as many component instances as required. This requires more general kinds of algorithms, some of which are also not complete.

There are advanced CSP techniques such as conditional and generative constraint solvers as described in [27] that handle the issues of constraint relaxation and an unbounded number of components being added at problem-solving time with use of a class mode. Further, some optimization is often required: not all valid configurations are acceptable or suitable [20, 71]. [27] treats these as preferences: the general solution is to produce a subset of all solutions for the user. Explanations for failure, as well as success, are also problematic for CPS techniques, but [27] describes some advanced technologies for those as well. However, there is another approach that starts with a more complex model than constraints and classes, but has a simpler solution for dynamic constraints and unbounded parts addition to a configuration, as well as preferences and explanations.

3.3 Generate and Test

The simplest type of configuration problem solving is to "generate and test". If all of the configuration conditions are expressed as constraints, then either the solving algorithm consists of some sort of CPS that tries values until consistency is achieved, or a specialized solver that somehow distinguishes among the constraints and assigns values to some variables first and then checks for consistency because it is known that this is efficient for the problem domain.

A more general type of algorithm distinguishes the constraints as goals and restrictions, as described in Sect. 2.4, and tries to satisfy some goal or subgoal (requirement-type constraint) and then test to see whether any restriction-type constraint is violated. The simplest systems select both a goal and a way to achieve it at random. A more sophisticated system allows the use of special knowledge both to select a goal and the first or next method to try in achieving it [50]. Using such knowledge to select methods allows at least local preferences to be easily considered. For instance, in building a boat, one may always prefer to start with wood as a building material, and try concrete last.

An immediate benefit of such reasoning is that goals may be used to explain component assignments. Such explanations may involve subgoaling. The choice of a propeller is a means of pushing the water, which is a subgoal of the goal of configuring a propulsion system, which is itself a subgoal of the general boat design. Further, most configuration solving techniques, including constraint satisfaction, may assign parts to a configuration at an intermediate stage in problem solving that are not needed in the final configuration. With goals, parts may be filtered out of the configuration if they have no valid goal associated with them.

Goals also allow for problem decomposition in that any method may decompose the goal into subgoals that are more easily solved. For instance, given the goal of building an automobile, a method may decompose it into the subgoals of building a drive train and a chassis, which can be pursued separately, but with their individual solutions connected by compatibility and aggregate constraints that restrict the candidate solutions.

Once a goal and a method of achieving it is determined, then the consequences of this design decision must be examined. Some systems consider all consequences to be new constraints and some [50, 71] distinguish between new subgoals that have to be achieved (new requirements) and constraint restrictions that may be violated because of the decision (new constraint violations). In [50], constraint violations can only be fixed by retracting some previous design decision: i.e., backtracking, or by relaxing a constraint.

One advantage of such a system is that irrelevant constraints do not have to be considered: once a decision has been made, a search, by proof for example, can be made to determine which, if any, constraints (restrictions) might be violated. There may be many constraints that are simply irrelevant for a particular search path that leads to a solution. Another advantage is that constraint violation resolution can be deferred until such resolution can be better informed [27, 50]. Restricting violation resolutions to backtracking also reduces the search space.

It is not usually known at goal creation time whether the goal can be achieved: more design decisions need to be made, possibly resulting in new components or ways in which they are assembled or connected. If all possible ways of achieving a goal fail, possibly because they violate constraints, then this is a "goal block" and some previous design decision may have to be undone. If such backtracking is not part of the algorithm, then it will not be complete: some answers may be missed.

Some algorithms allow higher-level goals, such as "decompose the task" or "find some combination of components to satisfy the user requirement of high-speed". The method of achieving a goal in a configuration problem is often to add a component to the configuration. Simple configuration systems only attempt to fulfill direct requests for components and only add components in order to do so, while also checking for constraints associated with the components.

Any such configuration solver has to check whether a constraint is violated after making such a design decision. Aggregate constraints, such as weight, may need to be checked. And a constraint may introduce not only new subgoals (such as connections or the need for more power) but also new constraints (such as no touching part may be made of iron).

When a constraint is violated, or a subgoal cannot be satisfied, then the system backtracks: some previous solution to a goal is retracted and an alternative way of satisfying it is tried. There is a wide range of backtracking methods available. Dependency-directed backtracking described above in Sect. 3.2, ensures that all possible solutions will eventually be found. This complete method is usually avoided since it can take a long time to find such a solution, unless special knowledge about the problem is used to determine the best fix [10, 14, 36, 50, 62].

A simpler technique is just to retract the last alternative tried: chronological backtracking. In the case that no special knowledge is used to fix a constraint violation, this method is simpler to implement and will do no worse than dependency-directed backtracking, though the latter may be able to employ special knowledge to make the search less expensive.

3.4 Specialized Generate and Test Algorithms

Because general CPSs and Generate and Test algorithms are expensive, more specialized special purpose algorithms are often used for specific types of configuration problems in which some knowledge about the problem domain can be exploited. Such specialized systems typically represent all conditions as constraints, but may distinguish some simple goals, such as the need to add some component or connection to the configuration (vs. say, the goal to decompose some configuration task).

These are usually implemented with some form of rule system. The two general types are "backward chaining" and "forward chaining", each implemented with engines of various efficiencies. The most common form of backward chaining rules systems use the Prolog [70] representation and the most common form of forward chaining systems use a Rete net [17] engine as they are very efficient in each case. (It is important to note that these are very restricted forms of logic that are not guaranteed to produce correct answers. There are modern systems using full First Order Logic that will produce provably correct answers in a reasonable time for an important set of problems.)

However, any configuration problem solving algorithm for any specific model may be programmed in either backward or forward chaining rule systems. They may be programmed in any programming language, for that matter. It is not necessary to use rules for an implementation and the choice of an implementation will depend upon the developers' familiarity with software tools as well as implementation efficiency.

The simplest type of configuration problem solver takes a list of components to be configured. The system takes the next component in the list and instantiates it. In this case, the instantiation is the goal and the way of achieving it is to create a component instance. Then the system looks to see if any new goals or restrictions have been created by this design decision. If there have been, then the goals/requests must be satisfied and the restriction violations must be removed, or else the instantiation design decision fails. If it fails, then the system must backtrack and try again in order to be complete. This can mean trying a different component or trying a different goal first.

Given a set of new goals and restrictions, the configuration system must try to satisfy them. A restriction has to satisfied only if it is violated. All new goals will need to be satisfied: these typically require the instantiation of one or more components, such as connectors, containers, or resource providers. A restriction violation might be of the form that the component must fit into a particular slot, which is already occupied. The only way to fix this constraint violation is to retract the most current decision to instantiate this component, or to retract the decision to place the current occupier of the slot there, or to retract the decision to instantiate the current slot occupier at all.

The simplest algorithm is only capable of creating yet another instance of the same component, because it only addresses specific component goals and takes each in order from an ordered list of component requests. Then there is no point in backtracking and the configuration attempt must fail.

Such simple component-based algorithms miss more solutions than others that are more flexible, such as algorithms that would allow goals that to be achieved by more than one component type, or possibly even combinations of them. However, they will find some solutions quickly, if they exist.

3.5 Case-Based Reasoning

Sometimes one would rather not solve the complete configuration but only change one that is similar. If one would like an automobile similar to one's neighbor, only with a bigger engine, then a reference to his existing configuration at the automobile dealer would be the right starting point.

However, often, the particular previous configuration is not known, so there must be a measure of similarity that is used to retrieve previous cases. The measure itself and the indexing of previous cases is problematic. There was extensive work done in this area, especially by the AG Richter in Kaiserslautern [45].

However, this technique is often a special case of using knowledge to guide the search, or of repairing an existing configuration, since the one retrieved does not fit the requirements. Since there was not an exact fit with requirements, and a little new configuration is done as possible, frequently the solutions range from conservative to poor [23], so we do not further discuss this approach here, but rather move on to specific systems that involve the reasoning methods of this chapter.

Chapter 4
Specific Configuration Systems

Abstract This chapter describes the development of twenty (20) automated configuration solvers, including both academic systems and patented systems from industry, using the models and reasoning techniques of previous chapters. Advances in the state of the art are discussed, as well as various advantages and disadvantages of the different systems.

4.1 Early Automated Configuration

By 1989, at least a dozen configuration systems using AI technologies had been reported in the literature, configuring computers, networks, operating systems, buildings, circuit boards, keyboards, printing presses, and trucks [21]. As previously discussed, all use some database, or catalog, of components, perhaps built-in to the algorithm code, but used stored externally to that code. Some of the earliest configured electrical circuits, such as VEXED [42]. Almost all used forms of test and generate algorithms. VEXED included no control knowledge and left it to the user to decide what next to test. The challenge is to select so that the configuration moves forward instead of backwards, removing previous selections: backtracking.

When testing against constraints failed, some form of backtracking was necessary, which could be very computationally expensive. The truth maintenance system (TMS) [16] was developed in response to the need to perform dependency-directed backtracking (DDB) [62] when constraints were violated and to avoid re-trying combinations of decisions that had already been tried.

However, undirected backtracking may result in too much search (though worst-case results seem rare), making some problems impossible to solve in a reasonable time. So very early, strategies were developed that could avoid backtracking. The commercial XCON/R1 system [39] used several strategies in a completely rule-based system that avoided backtracking in many if not most cases (backtracking was explicitly included in one subset of problem solving). One of these strategies

C. J. Petrie, *Automated Configuration Problem Solving*, SpringerBriefs in Computer Science, 23
DOI: 10.1007/978-1-4614-4532-6_4, © The Author(s) 2012

was to consider components as providing specific functions and to limit the space of possible configurations to known valid architectures.

The R1 system was very sophisticated and potentially solved all of the problems normally encountered in computer generation. This involved reasoning about physical structure (proximity, slots, ports, connecting cable lengths, and cabinet containment) as well as functional capabilities and resource allocations such as power.

The central problem of R1 turned out to be rule manageability. This is because many of the rules, and conditions embedded in rules, represented not knowledge about the configuration but rather knowledge about which rules should fire first in order to avoid backtracking. It is the embedding of such control knowledge that caused rule maintenance problems [40]. (This was also a common problem with Prolog and Rete-based reasoning systems.)

The next major development was the use of frames to represent the static knowledge about the components. Rules could then use this static knowledge as needed. When new information had to be added, then often only the frame representation had to be changed. This made rule management much easier.

One of the earliest attempts to overcome the weakness of the XCON/RI model was OCEAN [8]. This system used a component decomposition hierarchy and separated the components from the control knowledge. The specific hierarchy had subcomponents such as Input/Output with terminals, cages to contain cards, memory, and power supplies. The key innovation was the use of templates as a top-level user request.

These templates could be used to request specific components, component assemblies, and/or functions. The component catalog was searched to find matching parts and these were added and deleted as necessary to the configuration list. As determined by control strategies associated with each part assembly, the assemblies were decomposed and constraints checked. Some of the associated conditions are requirements such as a number of amperes for a power supply, which might need to be added to the configuration, or whether a particular CPU board was required. Other conditions associated with component assemblies would be restrictions, such as the maximum number of cards for memory. The search strategy was still similar to that of XCON/RI in that there was little backtracking and no particular backtracking strategy was described in the case of overall failure, suggesting that this algorithm was very incomplete: it may have missed a significant number of possible solutions.

COSSACK [20] was a general system in which components added requirement-type constraints that might introduce new components. Initial user requirements were cast as general constraints: there was no explicit distinction between requirement and restriction constraints. So for example, that one component must be connected to another, or a chip must be placed in a particular socket, or have any number of physical or electric properties were associated with particular components and introduced as that component was added to the configuration. A general hierarchical structure of components was used, including "Hardware Component" and subclasses such as "Connector", "Cage", "Slots", and "Ports" as in Fig. 2.1.

Generation of components was done to satisfy constraints, which were also used to test whether the current configuration was valid. Some control knowledge was domain-independent, such as attempting to find a single component that would satisfy multiple constraints. Backtracking was roughly chronological but the algorithm was not complete and would miss significant solutions, though more backtracking was possible than with R1.

From such specific configuration solving systems, there were several abstractions into general "shells" that could solve similar classes of problems, including DSPL from AIR-CYL [7], EDESYN from HI-RISE [35], and EVEXED from VEXED [42], MOLGEN [64], DONTE [66], and DESIGNER [30].

KEWB [33] was an improvement of COSSACK. The component-structured hierarchy was expressed in a (then) modern frame system, the representation language was improved, some constraints were represented explicitly as resources to be supplied and consumed, and most important, the backtracking was made more complete though inefficient when needed because no DDB was used.

Backtracking in KEWB was avoided often by associating constraints with component classes high in the structured hierarchy. If one of these constraints was violated, then that entire tree of possible components could be avoided initially.

However, none of these systems seemed appropriate for all configuration tasks and the problem of embedded control knowledge remained. One idea was to have rules about which rules to fire in a given situation [55]. However, this was too difficult to develop for most implementations.

4.2 Specialized Systems for Configuration

Simpler, specialized, systems just gave up trying to represent explicit control knowledge and would pick components at random and use chronological backtracking. This strongly limited the size and complexity of the configuration problems that could be attempted but was very effective for specific classes of problems, especially industrial problems. In the late 1980s and early 1990s, most dealt with the challenging but commercially important problem of computer configuration. Some of them focused on very specific sub-problems and at least one, Trilogy [18], focused on a very specific way of solving the more general computer configuration problem, discussed further below.

Cicciarelli and Millis [9] describes a semi-automatic method of configuration of cables for a given electronic device in conjunction with standard Computer Aided Design (CAD) techniques and standards. The basic configuration file in such a system is called a "wirelist", which is a CAD standard component database format. Components (cables) selected from an available set are added to and deleted from a temporary wirelist (and other files) until a desired configuration is achieved. The focus upon the correct physical location of connecting wires and their ports, as well as constraints on length and pin connections, is typical of CAD systems. There is little automatic reasoning in such a system and so backtracking is not specifically described.

Richek et al. [57] describes a very specific and complex semi-automatic (interactive) algorithm for configuring computer circuit boards. This technology concentrates on selecting boards and allocating physical slots for them, under the constraints of ports and memory resources. The algorithm checks whether an existing resource may be shared or more resources need to be allocated. There is a basic dependency-directed backtracking capability that uses a priority method of selecting which previous design decision should be retracted if necessary. The algorithm is not described well in the patent, but may be complete as the patent is explicit that there may be no possible solution in some cases.

Some early systems were successful because they adopted a very simple but general approach to configuration. NICAD [28] was used to configure plants, such as a soda recovery boiler for the paper industry. The components were modeled in the usual "is-a", "has-part" structured hierarchy using frames, and each slot of a component was associated with rules about how to compute values based upon the values of other components in the configuration. Changes in values of a component were propagated using these rules to values of other components. Such values included the number of instances of a particular component type, such as "support pillar".

NICAD essentially had no control knowledge and performed no backtracking as such. If a user made a selection at a higher-level component, this could cause a component lower in the hierarchy to be instantiated. However, changes in a lower level would propagate upwards, causing a kind of backtracking that had to be controlled by the user. The system was commercialized in 1989 and continued into the twenty-first century [44].

Trilogy [18] was more general than circuit board configuration but describes a very specific component structured hierarchy and associated with this component structured hierarchy a specific algorithm for solving computer configuration problems. This was an example specialized generate-and-test as discussed in Sect. 3.4 and was the basis for a very successful commercial system [5].

Because the hierarchy and algorithm is specific, it has the advantage of not having to represent much control knowledge that caused problems in earlier systems. The algorithm recurses on component requests, which may come from either a user request or a decomposition constraint, which is really a new requirement for components. Constraints are associated with components. Because of the specific hierarchy containing "container", "connection", the algorithm checks in order for constraints of container, connection, and then new component requirements. Connection requests take into account specific port constraints. The algorithm also checks for new resource requirements when these constraints are generated. The algorithm specifically checks to see whether new component and resource requests are already satisfied before attempting to add a new component to the configuration.

Backtracking is chronological in this system and the algorithm appears to be complete. Because the algorithm is so specific, no control knowledge is used, so this is in some sense, another "anti-knowledge" approach to configuration, except that it is the very specific component structured hierarchy model, a form of embedded implicit knowledge, that makes this algorithm possible. This system was very successful for the purpose of configuring computer systems and was widely used in Trilogy configuration products.

4.3 General Configuration Solvers

Several systems provide a more sophisticated model and algorithm for solving configurations than did the early systems. In the best cases, these provide some specialized representations for a kind of configuration in order to provide efficiency. An alternative is to provide a general framework in which models and algorithms for different problems can be implemented.

PLAKON [10] explicitly used a frame model for describing a component structure hierarchy with both "is-a" and "has-part" (component) relationships. Constraints are general relations associated with frame slot values, as well as the existence, or nonexistence of other components. In the case of the latter, the constraints are checked against the current configuration. If a required component already exists, nothing need be added. If some component is in the configuration but should not be, then backtracking is necessary.

There is no specific algorithm for PLAKON. Rather it allows for the representation of various control strategies. Notably, these strategies might contain branches and loops, which is a very difficult issue in planning a configuration. For backtracking, four modes are possible. The simplest is "XCON" mode, in which there is no backtracking. Chronological backtracking is facilitated, as is DDB. Parallel worlds were another mode proposed, using an ATMS [12], but this purely breath-first search method has shown not to be feasible for problems as described in [13]: the solution consists of a set of single choices of variable assignments that should be consistent, which characterizes most configuration problems. Some kind of backtracking is needed. PLAKON allowed for knowledge of how to fix a specific conflict.

COSMOS [25] took an elegant very simple but very general approach to configuration: resource balancing. COSMOS also used a frame system to express both the usual combination "is-a" and "has-part" component structure model, but also for all properties of such components expressed as resources, including connection properties, such as containing cages, slots and ports.

Components were generally selected to try to satisfy resource requests, beginning with one supplied by the user. The control knowledge represented was about which resources next to try to balance. Each component was described by the resources it consumed and/or provided. Resource balancing using such knowledge was the generation part of the algorithm, essentially requirement types of constraints. If a requested resource was already supplied, then balance was achieved without adding a new component.

Testing in the COSMOS algorithm was performed by matching the current partial configuration against restriction types of constraints. Backtracking was purely chronological. Due to the organization of resources associated with components, COSMOS was claimed to be easy to maintain and was used in several commercial applications, including Programmable Logic Controller configurations, taking into account price, power consumption, installation time, floor space, and connections.

Redux [50] was another general framework and engine for configuration, abstracted from the notion of goal/operator Hierarchical Task Network (HTN) planning [59, 65]. Redux was based upon the TMS and frame system from the earlier Proteus [46] system in which guided DDB [47] was successfully used in two commercial systems [63, 69]. A TMS works in Redux because there is a precise model [51] for constructing the TMS nodes that depends on more than just whether the nodes are antecedents or consequences of configuration rules, which overcomes the problems of the simpler uses of a TMS [48, 49].

Previous configuration systems sometimes mentioned two types of constraint conditions, requirements and restrictions, as discussed in Sect. 2.4, but they were not distinguished explicitly and often not treated differently by the solving algorithm as they are in Redux. Redux first distinguished them by calling requirements "goals" and restrictions "constraints": the former could only be satisfied, and the latter could only be violated, by making a design decision. As previously mentioned in Sect. 2.4, by first trying to satisfy goals, and then trying to resolve only those constraint conflicts that resulted, problem solving was simplified. Constraint relaxation is the final resort for problem solving.

It was further found useful to distinguish decision validity from optimality in a model of how to use a TMS for engineering configuration. Distinguishing validity from optimality allowed controlled propagation of changes. When the TMS changed the validity of an optimality node, it did not automatically change the validity of the relevant decision. For instance, if a flight was planned because it was cheaper, that part of the plan was not invalidated just because the price changed. In the previous systems, price was just another antecedent in the firing of the rule and the consequent, the choice of flight, would become invalid as soon as the price changed.

Explanations in Redux not only take advantage of the goal representation, but also the optimality reasons for decisions, including that a better decision may have been tried but needed to be revoked because of a constraint violation. The users will also receive notifications when such a violation no longer exists, so that this decision could be taken again. The Redux DDB produces the justifications to track such loss of *Pareto Optimality*.

Users are also notified when such revisions result in decisions that achieved now invalid goals, thus introducing unnecessary parts into the configuration. For example, the user may have decided to build a complex part out of components, and one subtask may be to machine the support frame. Later, the user may decide rather to use an off-the-shelf component with an integral support frame. The machining subtask is now invalid. The person assigned this task will be notified of this. Further, if the support frame had been made and assigned to the configuration, the person would be notified that the part should be removed. This is especially important in mass-produced objects, such as airplanes, where there is a necessity to stockpile an inventory for every part.

Like PLAKON, no specific search algorithm is provided other than blind search. With operators for types of goals and rules for choosing both, this "Goal-Operator-Oriented-Programming (GOOP)" technique can implement any test-and-generate configuration algorithm, including full planning for conjunctive goals [52]. Goals may be any configuration task required for problem solving, including, of course, task decomposition.

Rules are used to reason about which goal to be tried first, and which operator, resulting in a design decision, to be tried first. The last is a good way to represent preference knowledge. Redux will always try to prove the "best" operator to apply to the goal that has not yet been tried. Explanations can always be produced as to why some part of the configuration is currently valid, or has been invalidated.

When constraint violations resulted, the DDB [47] allows also rules to be used to reason about the fix. Thus control knowledge could be added as desired, and types of control knowledge distinguished. Control knowledge can also be used to defer constraint violation resolution until such resolution can be better informed [27]. With no control knowledge, the fundamental search algorithm is complete, assuming that the goals and operators are a complete model of the problem. The DDB will not allow thrashing: previously tried solutions will not be retried.

Several parts of the Redux system have been re-implemented for systems in both academia and industry. During the 1990s, several Master's and PhDs from the University of Kaiserslautern's AI Group, headed by Michael Richter, were based upon Redux, with work continuing at other universities [11, 38]. As some of these researchers moved into industry, such the former Daimler Research facility in Berlin, these ideas became more wide-spread. These ideas were also explored further at Stanford University [56], and used in private research projects with companies such as Hughes Aircraft, Toshiba, NEC, HP, and SAP.

Chapter 5
Final Notes on Configuration Solving

Abstract This conclusion reviews the lessons learned; outstanding issues; the relevance of academic research to industrial development; recent work in automated configuration problem solving such as SAT and Logical Spreadsheets; and the relevance of this field of research to modern problems, such as web service composition.

5.1 A Few Most Important Lessons Learned

The most important lesson is that is likely that configuration problems are in general not solvable efficiently with a single algorithm but configuration algorithms should rather be developed with a particular problem set in mind.

Differentiating constraints among goals and restrictions is particularly useful when the goals are not simply assignments of values to variables but can be achieved by more complicated reasoning involving subgoals. However, this distinction is also useful in interactive configuration.

For instance, in some versions of logical spreadsheets [31] that solve a multiple meeting/room assignment problem, it is useful to color blue those meetings that have yet to be assigned a room and should be, and to color red those meeting assignments already made that conflict with each other. The user can decide what to do about each kind of issue and when. The user can also ask the program to decide automatically: one presses the "Visine"(TM) button to get the red out and the "Zoloft"(TM) button to remove the blues.

Deferred resolution of constraint violation is especially useful when constraints are dynamic and hierarchical: that is, some choices of values for some variables impose new constraints or even choices of values for other variables, as determined by the configuration developer. The user may not want either the new constraints or new values, and need to somehow backtrack to the higher-level choices that led to these undesirable choices. This lesson has been proven a valuable one even in today's modern systems.

C. J. Petrie, *Automated Configuration Problem Solving*, SpringerBriefs in Computer Science,   31
DOI: 10.1007/978-1-4614-4532-6_5, © The Author(s) 2012

An interesting user interface problem is how to present the various choices for resolving the outstanding constraint violations.

5.2 Outstanding Issues

There are certainly outstanding computational issues still in configuration research [19]. As with most computer science problems, one is efficient scaling: larger problems are typically solved with increasingly specialized approaches.

Another is multiple technique integration. An important example is that test-and-generate together with constraint satisfaction could be very powerful, but dependency-directed backtracking is inconsistent with the latter [26], though combining some version of DDB with SAT has been done [37]. The integration is needed however to deal with such problems as sensitivity analysis and trade-offs. For example, when the plane as designed is too heavy, which parts of the plane should be lightened and what are the effects of such change?

Another research area is that of coordinating distributed configuration in parallel with artifact construction [56]. Extending the Internet to facilitate configuration-style coordination is one approach to the new area of coordination engineering [53].

5.3 Industrial Versus Academic Research and Development

This monograph covered both academic and commercial systems. The latter are not as well-known as they tend not to published. The former are not as useful in industry because there is no incentive to be so in academia and there is also no upper bound on complexity. Redux [50] is a good example. Even in academia, only parts and principles were reused as the entire system was too complex.

This is a good reason why many of the academic systems have not been taken up by industry, as noted by Guenter and Kuehn [23]. Even though solutions to hard industrial problems were demonstrated, the Redux technology, for example, was found too different from ones already in use for adaptation. This was also the case for the original Proteus use by NCR [63]. Even though it was used successfully by NCR's customers to design chips, the product had to be abandoned after a few years because of the lack of industry engineers and programmers who understood the underlying technology.

Beyond complexity, there is a more general reason why general knowledge-based systems are adapted by industry. In order to function well, the domain must be completely modeled, requiring labor-intensive knowledge engineering. However, such systems are better than typical hard-coded programming systems usually when the problems are not routine, which requires more knowledge engineering. Typically, some easier technology is used for specialized cases. However, academia makes an important contribution by understanding the general cases for which there are easier solutions in special cases.

5.4 More Recent Work

This monograph only reviews configuration solving up through the early 1990s as configuration has become a standard technology and has been specialized in many ways many times in the last two decades. Modern systems facilitate the user's navigation to such choices and aid in finding acceptable but consistent solutions efficiently. This review should serve as a basis for the student of configuration to understand the history of techniques for configuration problems solving. Most modern systems will find their basis in one of the historical systems described here.

Constraint satisfaction continues to be a leading technique for configuration as progress is made that field and in newer but related technologies as described in [27]. In particular, SAT [37, 61] has been widely used and seems to be incorporating some TMS techniques, including a version of DDB.

Newer work in computational logic can also be used in the form of Logical Spreadsheets [31] that use first-order logic to propagate changes in values through logical expressions and allow users to resolve conflicts when built-in resolution policies are not sufficient. Such systems can also be used as an alternative to DDB for deriving and presenting constraint violations [32]. This logic-based approach also facilitates novel methods of guiding the user through conflict resolution.

Much of the configuration system building energy now rests inside commercial companies performing configuration for specialized areas, especially sales [5]. An example of an important modern system is the very specialized configuration system for SAP's Business by Design(TM) system that requires the user to manage many constraints and dependencies among the system components being configured.

In academia, there are newer topics that are essentially specializations of the configuration problem as well, so that this kind of problem continues to be important. The composition of web services has been an important topic in computer science in the first part of the twenty-first century. All of the problems posed in the literature are essentially a configuration problem as there are a finite set of types of services at any given time, even though they may have to be discovered. However, no new services are developed in order to solve the problem. Configuration-type planning is sufficient to solve all such problems though simple goal regression is sufficient for most cases [52]. The crucial problem yet to be solved in service composition is not that of configuration but that of standardizing the terms in the distributed service descriptions [54].

Web services composition is but example of how the fundamental technologies for configuration problem solving developed in the last century continue to be relevant and why modern researchers would do well to become familiar with them.

References

1. Abelson H, Sussman J (1985) The structure and interpretation of computer programs. MIT Press, Cambridge
2. Baginsky W, Endres H, Geissing G, Philipp L (1988) Basic architectural features of configuration expert systems for automatic engineering. In: Proceedigs of Internet Workshop on AI for Industrial Applications, IEEE Press, Piscataway 603–607
3. Baykan C, Fox M (1991) Constraint satisfaction techniques for spatial planning. In: Proceedigs of Intelligent CAD systems III, Springer-Verlag New York, Inc. New York
4. Birmingham WP, Gupta AP, Siewiorek DP (1992) Automating the design of computer systems: The MICON Project, Jones and Bartlett
5. Bois R (2007) Sales configurationfrom efficiency to excellence, AMR Research, http://www.bigmachines.com/downloads/amr research07.pdf. Last Accessed 4 April 2012
6. Brown D, Chandrasekraran B (1989) Design problem solving: knowledge structures and control strategies, Research Notes in Artificial Intelligence, Pitman, London
7. Brown D (1998) Defining configuring, AI EDAM special issue on Configuration. Darr, McGuinness, Klein (eds.) Cambridge University Press, Cambridge http://www.cs.wpi.edu/dcb/Config/EdamConfig.html
8. Bennet S, Lark J (1986) US Patent 4591983
9. Cicciarelli R, Millis D (1989) US Patent 4870591
10. Cunis R, Guenter A, Syska I, Peters H, Bode H (1989) PLAKON- An approach to domainindependent construction. In: Proceedigs of 2nd international conference on Industrial applications of artificial intelligence and expert systems vol 2, ACM, New York
11. Dellen B, Kohler K, Maurer F (1996) Integrating software process models and design rationales. In: Proceedigs of The 11th Knowledge-Based Software Engineering Conference, IEEE Press, Piscataway, 84–93
12. DeKleer J (1986) An Assumption-based TMS, Journal of AI, 28(2):127–162, AAAI Press, Menlo Park
13. DeKleer J, Williams B (1986) Back to backtracking, AAAI-86, pp 910–917, AAAI Press, Menlo Park
14. Dhar V et al. (1985) An approach to dependency-directed backtracking using domain specific knowledge, NYU technical report IS-85-21, New York
15. Feigenbaum E, McCorduck P (1983) The fifth generation (1st ed.), Addison-Wesley, Reading
16. Doyle J, A Truth Maintenance System (1979) Artificial Intelligence, 12(3):251–272, AAAI Press, Menlo Park
17. Forgy, CL (1982) Rete: a fast algorithm for the many pattern/many object match problem, Artif. Intell. 19(1):17:37 Morgan Kaufman, San Francisco
18. Franke D, Lynch J (1999) US Patent 6002854

19. Franke D (1998) Configuration research and commercial solutions, AI EDAM 12:295-300, Cambridge University Press, Cambridge

20. Frayman F, Mittal S (1987) COSSACK: a constraints-based expert system for configuration tasks, Knowledge Based Expert Systems in Engineering: Planning and Design, Sriram S and Adey C (eds.), 143–166, Computational Mechanics Publications, Portsmouth

21. Frayman F, Mittal S (1989) Towards a generic model of configuration tasks. Proc. 11th International Joint Conference Artificial Intelligence, 1395–1401, Morgan Kaufman, San Francisco

22. Genesereth MR, Nilsson NJ (1987) Logical foundations of artificial intelligence, Morgan-Kaufman, Waltham

23. Guenter A, Kuehn C (1999) Knowledge-based configuration- survey and future directions, XPS-99: Knowledge based systems, Puppe F (ed), 1570:47–66, Springer, Berlin. doi: 10.1007/10703016-3

24. Haag A (1998) Sales configuration in business processes, IEEE Intelligent Systems, 13(4):7885, IEEE Press, Piscataway

25. Heinrich M, Juengst EW (1991) A resource-based paradigm for the configuring of technical systems from modular components, 257–263, IEEE Press, Piscataway

26. Jeon H, Petrie C, Cutkosky M (1997) Combining constraint propagation and backtracking for distributed engineering, AAAI'97 Workshop on Constraints and Agents, Technical Report WS-97–05, AAAI Press, Menlo Park

27. Junker U (2006) Configuration, Handbook of constraint programming, Rossi F, van Beek P, Walsh T (eds), 837–873, Elsevier, New York

28. Karonen O (1987) Intelligent CAD, Norweian AI Society, (NAIS) Seminar

29. Laurent JP, Ayel J, Thome F, Ziebelin D (1984) Comparative evaluation of three expert system development tools: Kee, Knowledge Craft, Art, The Knowledge Engineering Review 1:18–29 Cambridge University Press,Cambridge DOI: 10.1017/S0269888900000631

30. Kant E (1985) Understanding and automating algorithm design, Trans. on Software Engineering, SE-11:(11)1361–1374, IEEE Press, Piscataway

31. Kassoff M, Zen L, Garg A, Genesereth MR (2005) Predicalc: a logical spreadsheet management system, 31st International Conference on Very Large Databases (VLDB), 1247–1250, Trondheim, Norway

32. Kao E, Genesereth MR (2011) Incremental Consequence Finding in First-Order Logic. Technical Report, Stanford University LG-2011–01

33. Kramer B (1991) Knowledge-based configuration of computer systems using hierarchical partial choice. IEEE International Conference on Tools for AI, IEEE Press, San Jose DOI 10.1109/TAI.1991.167117

34. Kumar V (1992) Algorithms for constraint satisfaction problems: a survey, AI Magazine 13(1)32-44, AAAI Press, Menlo Park

35. Maher ML (1988) HI-RISE: an expert system for preliminary structural design, Expert Systems for Engineering Design, M. D. Rychener (Ed), 37–52, Academic Press, Inc, San Diego

36. Marcus S, Stout J, McDermott J (1987) VT: an expert elevator designer that uses knowledgebased backtracking, AI Magazine, 8(4):41–58, AAAI Press, Menlo Park

37. Marques-Silva JP, Sakallah K, Marques JP, Karem S, Sakallah A (1996) Conflict analysis in search algorithms for propositional satisfiability. In: Proceedigs of the IEEE International Conference on Tools with Artificial Intelligence, pp 467–469, IEEE Press doi: 10.1109/TAI.1996.560789

38. Maurer F, Dellen B, Bendeck F, Goldmann S, Holz H Kötting K (2000) Project planning and web-enabled dynamic workflow technologies, IEEE Internet Computing, 4(3):65–74, IEEE Press

39. McDermott J (1982) R1: A rule-based configurer of computer systems, Artif. Intell., 19(1): 39–88, Morgan Kaufmann, San Francisco

40. McDermott J (1984) R1 Revisited: four years in the trenches, AI Magazine, 5(3), AI Press, Menlo Park
41. McGuinness D,Wright J (1998) Conceptual modelling for configuration: a description logicbased approach, Artificial Intelligence for Engineering Design, Analysis and Manufacturing, 12(4):333344, Academic Press, Inc, Academic Press, Inc, San Diego
42. Mitchell T et al. (1983) An intelligent aid for circuit redesign, Proc. AAAI, 274-278, AAAI Press, Menlo Park
43. Mostow J (1985) Toward better models of the design process, AI Magazine, 6(1):44–57, AAAI Press, Menlo Park
44. Nurminen J et al. (2003) What makes expert systems survive, Expert Systems with Applications, 24:199–211, Pergamon Press (Elsevier)
45. Paulokat J, We S (1993) Fallauswahl und fallbasierte Steuerung bei der nichtlinearen hierarchischen Planung, Beitrge zum 7. Workshop Planen und Konfigurieren, Hamburg
46. Petrie C (1985) Proteus—a default reasoning perspective, MCC Technical Report TR AI-352–86
47. Petrie C (1987) Dependency-directed backtracking revised for default reasoning, Proc. AAAI-87, 167–172, AAAI Press, Menlo Park
48. Petrie C (1989) Reason maintenance in expert systems, Kuenstliche intelligenz,2/89:54–60. Also MCC TR ACA-AI-021-89
49. Petrie C (1989) Using a TMS for problem formulation (1989) Proc. IJCAI-89 Workshop on Constraint Processing, Cognitive Systems Lab, UCLA, 1989. AlsoMCC TR ACT-AI-341-89
50. Petrie C (1991) Context Maintenance, Proc. AAAI-91, 1:288–295, AAAI Press, Menlo Park
51. Petrie C (1993) The redux' server (1993) Proc. Internat. Conf. on Intelligent and Cooperative Information Systems (ICICIS), Rotterdam. Also MCC TR EID-001-93
52. Petrie C (2009) Planning process instances with web services. In: Proceedigs of ICEIS AT4WS, 31–41, SciTePress, Milan, Italy. dx.doi.org/10.5220/0002171400310041
53. Petrie C (2011) Enterprise coordination on the internet, Future Internet 3(1):49–66; doi: 10.3390/fi3010049
54. Petrie C, Hochstein A, Genesereth MR (2011) Semantics for smart services, The science of service systems, Demirkan H, Spohrer JC, Krishna V (eds) Springer, New York
55. Petrie C, Huhns M (1988) Controlling forward rule instances, Proc. 8th Internat. Workshop Expert Systems and Applications, 383-398, Avignon, IEEE Press, Piscataway
56. Petrie C, Goldmann S, Raquet A (1999) Agent-based project management, Lecture Notes in AI 1600:339 Springer-Verlag, Berlin/Heidelberg. DOI: 10.1007/3-540-48317-9-14
57. Richek M, Gready R, Jones C (1993) US Patent 5257387
58. Rossi F, Dhar V, Petrie C (1990) A new notion of csp equivalence, Proc. ECAI-90:550-556, Stockholm, Sweden. Also available as the longer MCC Technical Report TR AI-022-89: On the equivalence of constraint satisfaction problems
59. Sacerdoti ED (1975) The nonlinear nature of plans, Proc. 4th Int. Joint Conf. on Artificial Intelligence (IJCAI), 206–214. Morgan Kaufmann, San Francisco
60. Schreiner AT (1993). Object oriented programming with ANSI-C. Hanser Verlag. ISBN 3- 446-17426-5. hdl:1850/8544
61. Sinz C, Kaiser A, KuchlinW(2003) Formal methods for the validation of automotive product configuration data. Artificial Intelligence for Engineering Design, Analysis and Manufacturing, 17(1):7597, Academic Press, Inc
62. Stallman R, Sussman G (1987) Forward reasoning and dependency-directed backtracking in a system for computer aided circuit analysis, Artificial Intelligence, 9(2):135–196, Morgan Kaufmann, San Francisco
63. Steele R, Richardson S, Winchell M (1989) Designadvisor: a knowledge-based integrated circuit design critic, Proc. IAAI-89, 135–142, AAAI Press
64. Stefik M (1980) Planning with constraints (MOLGEN: Part 1 and Part 2), Artif. Intell. 16(2):111–169, Morgan Kaufmann, San Francisco

65. Tate A (1977) Generating project networks, Proc. 5th Int. Joint Conf. on Artificial Intelligence (IJCAI), 888–893. Morgan Kaufmann

66. Tong C (1990) Knowledge-based design as an engineering science: the Rutgers AI/Design Project, in: Applications of AI in Engineering V, Vol. 1: Design, Gero JS (Ed.). In: Proceedigs of 5th International Conference on Applications of AI in Eng., 297–319, Boston, MA, Computational Mechanics Publications, Springer-Verlag, Berlin

67. Tong D (1988) US Patent 4896269

68. VanMell W (1980) A domain-independent systemt that aids in constructing knowledge-based consultation programs. Report STAN-CS-80-820, Departmentof Computer Science,Stanford University

69. Virdhagriswaran S, Levine S, Fast S, Pitts S (1987) PLEX: a knowledge based placement program for printed wire boards, 302–305. Third IEEE Conference on AI Applications, IEEE Press, Orlando

70. Warren DHD, Prolog: the language and its implementation compared with Lisp (1977) In: Proceedigs of the 1977 symposium on AI and programming languages, ACM, New York. DOI 10.1145/800228.806939

71. Wiellinga B, Schreiber G (1997) Configuration-design problem solving, IEEE Expert, 12(2):49–56, IEEE Press